Fuzzy Inventory Model for Deteriorating Item

Authors Details: Dr. Harish Nagar & Priyanka Surana

CONTENTS

Chapters	Page No.
Abstract	3
1. Introduction to the topic of study	4-23
2. Definitions and Notation	24-31
3. Mathematical model and Numerical example	32-57
4. .Conclusion	58
References	59-62

Abstract:

While the development and application of fuzzy inventory control models of deteriorating item is one of the main concerns of researchers and practitioners, most studies done in this field till trapezoidal fuzzy numbers by using different method. In this dissertation at first, we study research done on PFN by signed distance method and graded mean representation method to defuzzify the total cost function. Fuzziness is introduced by allowing the cost components (holding cost, shortage cost, etc.), demand rate and the deterioration. In which demand increase with time and shortage are fully backlogged. Then by, comparison of developed models some theoretical and practical results are derived, and various directions are suggested for future research.

Chapter-1
Introduction to the topic of study

1.1 Fuzzy related definitions and its applications.

1.2 Fuzzy inventory control system

1.3 Deteriorating item

1.4 History

1. Introduction to the topic of study

You have to crawl before you can fly, so we're going to ease into the Fuzzy World Tour with some very elementary fuzzy definitions. The real world is up and down, constantly moving and changing, and full of surprises .In other word, Fuzzy. Fuzzy techniques let you successfully handle real world situations.

In every day content most of the problems involve imprecise concept. To handle the imprecise concept, the conventional method of set theory and numbers are insufficient and need to be extended to some other concepts. Fuzzy concept is one of the concepts for this purpose.

1.1-Fuzzy related definitions and its applications:-

A **fuzzy concept** is a concept of which the boundaries of application can vary considerably according to context or conditions, instead of being fixed once and for all [25]. This means the concept is vague in some way, lacking a fixed, precise meaning, without however being unclear or meaningless altogether [20]. It has a definite meaning, which can become more precise only through further elaboration and specification, including a closer definition of the context in which the concept is used.

A fuzzy concept is understood by scientists as a concept

Which is "to an extent applicable" in a situation, and it therefore implies gradations of meaning. The best-known example of a fuzzy concept around the world is an amber traffic light, and indeed fuzzy concepts are nowadays widely used in traffic control systems [23].

Fuzzy concepts can generate uncertainty because they are imprecise (especially if they refer to a process in motion or a process of transformation where something is "in the process of turning into something else"). In that case, they do not provide a clear orientation for action or decision-making ("what does X really mean or imply?");

reducing fuzziness, perhaps by applying fuzzy logic, would generate more certainty. However, this is not necessarily always so A concept, even although it is not fuzzy at all, and even though it is very exact, could equally well fail to capture the meaning of something adequately. That is, a concept can be very precise and exact, but not - or insufficiently - applicable or relevant in the situation to which it refers. In this sense, a definition can be "very precise", but "miss the point" altogether.

A fuzzy concept may indeed provide more security, because it provides a meaning for something when an exact concept is unavailable - which is better than not being able to denote it at all. A concept such as God, although not easily definable, for instance can provide security to the believer.

Fuzzy concepts can be used deliberately to create ambiguity and vagueness, as an evasive tactic, or to bridge what would otherwise be immediately recognized as a contradiction of terms. They might be used to indicate that there is a connection between two things, without giving a complete specification of what the connection is, for some or other reason. This could be due to a failure or refusal to be more precise. But it could also be a prologue to a more exact formulation of a concept, or a better understanding logic. The novelty of fuzzy logic is, that it "breaks with the traditional principle that formalisation should correct and avoid, but not compromise with, vagueness".

Fuzzy set-

In mathematics, **fuzzy sets** are setting whose elements have degrees of membership. Fuzzy sets were introduced by Lotfi A. Zadeh [15] and Dieter Klaua [14] in 1965 as an extension of the classical notion of set. Different types of fuzzy sets are defined to clear the vagueness of the existing problems. Membership function of these sets, which have the form $A: R \rightarrow [0, 1]$ and it has a quantitative number called fuzzy number

[10] .At the same time, Salii (1965) defined a more general kind of structures called L-relations, which were studied by him in an abstract algebraic context. Fuzzy relations, which are used now in different areas, such as linguistics (De Cock, et al., 2000), decision-making (Kuzmin, 1982) and clustering (Bezdek, 1978), are special cases of L-relations when L is the unit interval [0, 1].

Example of fuzzy set:-

Example 1:-We consider statement "Jenny is young". At this time, the term "young" is vague. To represent the meaning of "vague" exactly, it would be necessary to define its membership function. When we refer "young", there might be age which lies in the range [0, 80] and we can account these "young age" in these scope as a continuous set. The horizontal axis shows age and the vertical one means the numerical value of membership function. The line shows possibility (value of membership function) of being contained in the fuzzy set "young". For example, if we follow the definition of "young" as, ten-year-old boy may well be young. So, the possibility for the "age ten" to join the fuzzy set of "young is 1. Also, that of "age twenty-seven" is 0.9. But we might not say young to a person who is over sixty and the possibility of this case is 0. Now we can manipulate our last sentence to "Jenny is very young". To be included in the set of "very young", the age should be lowered. If we define fuzzy set as such, only the person who is under forty years old can be included in the set of "very young". Now the possibility of twenty-seven-year-old man to be included in this set is 0.5.

That is, if we denote A= "young" and B="very young",

$$\mu_A(27) = 0.9, \mu_B(27) = 0.5.$$

Example 2-What's the process of parallel parking a car? [7]

First you line up your car next to the one in front of your space. Then you angle the car back into the space, turning the steering wheel slightly to adjust your angle as you get closer to the curb. Now turn the wheel to back up straight and—nothing. Your rear tire's wedged against the curb. OK. Go forward slowly, steering toward the curb until the rear tire straightens out .Fine except, you're too far from the curb. Drive back and forth again, using shallower angles .Now straight forward. Good, but a little too close to the car ahead. Back up a few inches. Oops, that's the bumper of the car in back. Forward just a few inches. Stop! Perfect!! Congratulations. You've just parallel-parked your car.

And you've just performed a series of fuzzy operations. Not fuzzy in the sense of being confused. But fuzzy in the real-world sense, like "going forward slowly" or "a bit hungry" or "partly cloudy"—the distinctions that people use in decision-making all the time, but that computers and other advanced technology haven't been able to handle.

Do all these situations have something in common? For one thing, they're all complex and dynamic. Also, like parallel parking, they're more easily characterized by words and shades of meaning than by mathematics.

In this dissertation you'll be immersed in the fuzzy world, not an easy process. You'll meet the basics, manipulate the tools (simple and complex), and use them to solve real-world problems .i.e. The real world is up and down, constantly moving and changing, and full of surprises .In other words, fuzzy. Fuzzy techniques let you successfully handle real world situations.

Characteristics of fuzziness:

•Word based, not number based. For instance, hot; not 85^0C.

• Nonlinear changeable.

• Analog (ambiguous), not digital (Yes/No).

Fuzzy interval:-

A fuzzy interval is an uncertain set $\tilde{A} \subseteq R$ with a mean interval whose elements possess the membership function value $\mu_A(x) = 1$. As in fuzzy numbers, the membership function must be convex, normalized and at least segmentally continuous. [1]

Fuzzy logic:-

A logic that is not very precise is called a fuzzy logic. The imprecise way of looking at things and manipulating them is much more powerful than precise way of looking at them and then manipulating them. Fuzzy logic is one of the tools for making computer system capable of solving problems involving imprecision. Fuzzy logic is an attempt to capture imprecision by generalizing the concept of set to fuzzy set.

The basic idea of fuzzy logic is, that a real number is assigned to each statement written in a language, within a range from 0 to 1, where 1 means that the statement is completely true, and 0 means that the statement is completely false, while values less than 1 but greater than 0 represent that the statements are "partly true", to a given, quantifiable extent. Susan Haack comments: "Whereas in classical set theory an object either is or is not a member of a given set, in fuzzy set theory membership is a matter of degree; the degree of membership of an object in a fuzzy set is represented by some real number between 0 and 1, with 0 denoting *no* membership and 1 *full* membership."[26] "Truth" in this mathematical context usually means simply that "something is the case" or that "something is applicable". This makes it possible to analyze a distribution of statements for their truth-content, identify data patterns, make inferences and predictions, and model how processes operate. Fuzzy logic in principle allows us to give a definite, precise answer to the question: "to what extent is something the case?", or "to what extent is something applicable?". Via a series of switches, this kind of reasoning can be built into electronic devices. That was already happening before fuzzy logic was invented, but using fuzzy logic in modelling has become an important aid in design, which creates many new technical possibilities.

Fuzzy number:-.

This can be likened to the funfair game "guess your weight," where someone guesses the contestant's weight, with closer guesses being more correct, and where the guesser "wins" if he or she guesses near enough to the contestant's weight, with the actual weight being completely correct (mapping to 1 by the membership function).

Fuzzy set theory:

In 1965, L. A. Zadeh introduced the concept of fuzzy set theory. Fuzzy set theory is an extension of classical set theory.

In classical set theory, the membership of elements in a set is assessed in binary terms according to a bivalent condition — an element either belongs or does not belong to the set. By contrast, fuzzy set theory permits the gradual assessment of the membership of elements in a set; this is described with the aid of a membership function valued in the real unit interval [0, 1]. Fuzzy sets generalize classical sets, since the indicator functions of classical sets are special cases of the membership functions of fuzzy sets, if the latter only take values 0 or 1.[3] In fuzzy set theory, classical bivalent sets are usually called crisp sets. The fuzzy set theory can be used in a wide range of domains in which information is incomplete or imprecise, such as bioinformatics.[16]

The fuzzy set theory is developed for solving the phenomenon of fuzziness prevalent in the real world. Up to this point, the fuzzy set theory has been widely applied in many fields, such as applied science, medicine and inventory management. The application of fuzzy set concepts in inventory models have been proposed by many researchers. Pertrovic and Sweeney [6] fuzzified the demand, lead time and inventory level into triangular fuzzy numbers in an inventory control model, and then determined the order quantity with the fuzzy proposition's method. Yao *et al.* [17] investigated the Economic Lot Scheduling Problem (ELSP) with fuzzy demands. They used the 'Independent Solution' as well as the 'Common Cycle' approach to solve the fuzzy ELSP problem. Yao *et al.* [12] presented a fuzzy inventory system without the backorder model in which both the order quantity and the total demand were fuzzified as the triangular fuzzy numbers. Chang [8] discussed the Economic Order Quantity (EOQ) model with imperfect quality items by applying the fuzzy sets theory and proposed the model with both a fuzzy defective rate and a fuzzy annual demand. Chang *et al.* [9] considered the mixture inventory model involving variable lead time with backorders and lost sales. They fuzzified the random lead-time demand to be a fuzzy random variable and the total demand to be the triangular fuzzy number. Based on the centroid method of defuzzification, they derived an estimate of the total cost in the fuzzy sense. Chen *et al.* [22] introduced a fuzzy economic production quantity model with defective products in which they considered a fuzzy opportunity cost, trapezoidal fuzzy cost and quantities in the context of the traditional production inventory model. Maiti [33] developed a multi-item inventory model with stock-dependent demand and two-storage facilities in a fuzzy environment (where purchase cost, investment amount and storehouse capacity are imprecise) under inflation and incorporating the time value of money.

Crisp and fuzzy relation: - One of the most fundamental notions in pure and applied sciences is the concept of a relation. Science has been described as the discovery of relations between objects, states and events. Fuzzy relations generalize the concept of relations in the same manner as fuzzy sets generalize the fundamental idea of sets.

Crisp Relation:- Crisp relation is defined on the Cartesian product of two universal sets determined as

$$X \times Y = \{(x,y) | x \in X, y \in Y\}$$

The crisp relation R is defined by its membership function

$$\mu_R(x,y) = \begin{cases} 1, (x,y) \in R \\ 0, (x,y) \notin R \end{cases}$$

Here "1" implies complete truth degree for the pair to be in relation and "0" implies no relation.

When the sets are finite the relation is represented by a matrix R called a relation matrix

If a crisp relation R represents that of from sets A to B, for $x \in A$ and $y \in BB$, its membership function $\mu_R(x,y)$ is,

$$\mu_R(x,y) = \begin{cases} 1, (x,y) \in R \\ 0, (x,y) \notin R \end{cases}$$

This membership function maps $A \times B$ to set $\{0,1\}$.

$$\mu_R : A \times B \to \{0,1\}$$

We know that the relation R is considered as a set. Recalling the previous fuzzy concept, we can define ambiguous relation.

Fuzzy relation:-

The fuzzy relation equation is an equation of the form A · R = B, where A and B are fuzzy sets, R is a fuzzy relation, and A · R stands for the composition of A with R.

A relation is a mathematical description of a situation where certain elements of sets are related to one another in some way. Fuzzy relations are significant concepts in fuzzy theory and have been widely used in many fields such as fuzzy clustering, fuzzy control and uncertainty reasoning. They also play an important role in fuzzy diagnosis and fuzzy modelling. When fuzzy relations are used in practice, how to estimate and compare them is a significant problem. Uncertainty measurements of fuzzy relations have been done by some researchers. Similarity measurement of uncertainty was introduced by Yager who also discussed its application.

Fuzzy relations generalize the concept of fuzzy sets to multidimensional universes and introduce the notion of association degree between the elements of some universe of discourse. Fuzzy relations generalize the concept of relations in the same manner as fuzzy sets generalize the fundamental idea of sets. Operations with fuzzy relations are important to process fuzzy models constructed via fuzzy relations. Relations are associations and remain at the very basis of most methodological approaches of science and engineering. Fuzzy relations are more general constructs than functions; they allow dependencies between several variables to be captured without necessarily committing to any particular directional association of the variables being involved.

Fuzzy relation has degree of membership whose value lies in [0, 1].

$$\mu_R : A \times B \to \{0,1\}$$

$$R = \{((x,y), \mu_R(x,y) \,/\, \mu_R(x,y) \geq 0, x \in A \text{ and } y \in B)\}$$

Application:-

In **philosophical logic**, fuzzy concepts are often regarded as concepts which in their application, or formally speaking, are neither completely true nor completely false, or which are partly true and partly false; they are ideas which require further elaboration, specification or qualification to understand their applicability (the conditions under which they truly make sense).

In **mathematics and statistics**, a fuzzy variable (such as "the temperature", "hot" or "cold") is a value which could lie in a probable range defined by quantitative limits or parameters, and which can be usefully described with imprecise categories (such as "high", "medium" or "low") using some kind of qualitative scale.

In **mathematics and computer science**, the gradations of applicable meaning of a fuzzy concept are described in terms of *quantitative* relationships defined by logical operators. Such an approach is sometimes called "degree theoretic semantics" by logicians and philosophers, but the more usual term is fuzzy logic or many-valued.[11] Fuzzy concepts often play a role in the creative process of forming new concepts to understand something. In the most primitive sense, this can be observed in infants who, through practical experience, learn to identify, distinguish and generalise the correct application of a concept, and relate it to other concepts. However, fuzzy concepts may also occur in scientific, journalistic, programming and philosophical activity, when a thinker is in the process of clarifying and defining a newly emerging concept which is based on distinctions which, for one reason or another, cannot (yet) be more exactly specified or validated. Fuzzy concepts are often used to denote complex phenomena, or to describe something which is developing and changing, which might involve shedding some old meanings and acquiring new ones.

1.2 Fuzzy inventory control system:-

In our daily life, we observe that a small retailer knows roughly the demand of his customers in a month or a week and accordingly places orders on the wholesaler to meet the demand of his customers .but this is not the case with a manager of a big departmental store or a big retailer ,because the stocking in such cases depends upon various factors ,e.g. demand ,time of ordering, lag between orders and actual receipts, etc. so the real problem is to have a compromising between over stocking and under stocking. Mathematically, the problem of maintaining the inventory arises due to the fact that ---if a person (e.g. big retailer) decides to have a large stock, his holding cost C_1 increases but his shortage cost C_2 and set up cost C_3 decreases. On the other hand, we are aware of the fact that the inventory is maintained for efficient and smooth running of business affairs. The study of such type of problem is known by the term "inventory control".

Inventory is needed to allow balancing between supply and demand. Careful inventory control is needed to make good economic sense. Considering demand uncertainty is important in inventory control.

Inventory control system: - An inventory control system is a system that monitors the number of items and location of those items in a warehouse or throughout a supply chain. They can be very simple or very complex.

In other words, the inventory control defines how often the stock level is reviewed to determine when and how much to order. It is performed on either a continuous or periodic basic types. In a continuous inventory control system, an order is placed for the same constant amount whenever the inventory on hand decreased to a certain level, whereas in a periodic system, an order is placed for a variable amount after the specific regular time interval [27]. Normally, a continuous system is used for Class A items, which represent a large percentage of the total dollar value of inventory. These inventory levels should be as low as possible, and safety stocks minimized. This requires accurate demand and supply estimations

Fuzzy inventory control model:- In crisp inventory models, all the parameters in the total cost are known and have definite values without ambiguity; in addition, the real variable of the total cost is positive. But, in reality, it is not so certain. Hence there is a need to consider the fuzzy inventory models. Due to the various fuzzy cases, one may consider different fuzzy inventory models as follows. Yao et al, Discussed fuzzy inventory with and without backorder models. Paper related to this paper treated fuzzy inventory with backorder. In [13], they fuzzified the shortage quantity s as a triangular fuzzy number in the total cost of inventory with backorder and kept the order quantity q as a crisp real variable. In this way, they obtained a fuzzy total cost. In[13] they fuzzified the order quantity q as triangular fuzzy numbers and trapezoid fuzzy numbers and kept the shortage quantity s as a crisp real variable in the total cost of inventory with backorder. In the authors used the extension principle to find the membership functions of the fuzzy total cost. Then they defuzzified by the centroid to find the estimate of the total cost in the fuzzy sense. Such methods are very difficult and complicated. In [24], they fuzzified the total demand quantity to an interval-valued fuzzy set in the total cost of inventory with backorder.

Then they used the extension principle to find the membership function and defuzzified to get an estimate of the total cost in the fuzzy sense. Papers, discussed fuzzy inventory without backorder.

In [13], they fuzzified order quantity q in the total cost of inventory without backorder to a triangular fuzzy number and trapezoid fuzzy number to get the fuzzy total cost. Then they used the extension principle to find their membership function. In [13], they fuzzified order quantity q and total demand quantity r in the total cost of inventory without backorder to triangular fuzzy numbers. In this way, they could compute a fuzzy total cost. Similarly, they used the extension principle to find their membership function. All the articles used the extension principle and centroid to find the estimate of the total cost in the fuzzy sense. This treatment is difficult and very complicated. Petrovic and Sweeney fuzzified the demand, lead time and inventory level into triangular fuzzy numbers in an inventory control model, then decided the order quantity by the method of fuzzy proposition. Vujosevic *et al.* fuzzified the ordering cost into a trapezoidal fuzzy number in the total cost of an inventory without backorder model and obtained the fuzzy total cost. They did the defuzzification by using centroid and obtained the total cost in the fuzzy sense. Chen *et al.* fuzzified the order cost, inventory cost, and backorder cost into trapezoidal fuzzy numbers and used the functional principle to obtain the estimate of the total cost in the fuzzy sense. Roy and Maiti rewrote the problem of classic economic order quantity into a form of nonlinear programming problem and introduced fuzziness both in the objective function and storage area. It was solved by fuzzifying both nonlinear and geometric programming techniques for linear membership functions. Ishii and Konno fuzzified the shortage cost L to a fuzzy number in a classical newsboy problem aimed at finding an optimal ordering quantity in the sense of fuzzy ordering.[13]

1.3 Deteriorating item:-

The definition of the word "deteriorated" would be decay, to grow worse, to weaken or to degenerate. An example of the word in a sentence could be, "his health deteriorated overnight".

Deteriorating items are common in our daily life; however, academia has not reached a consensus on the definition of the deteriorating items. According to the study of Wee HM in 1993, deteriorating items refers to the items that become decayed, damaged, evaporative, expired, invalid, devaluation and so on through time. According to the definition, deteriorating items can be classified into two categories. The first category refers to the items that become decayed, damaged, evaporative, or expired through time, like meat, vegetables, fruit, medicine, flowers, film and so on; the other category refers to the items that lose part or total value through time because of new technology or the introduction of alternatives, like computer chips, mobile phones, fashion and seasonal goods, and so on. Both of the two categories have the characteristic of short life cycle. For the first category, the items have a short natural life cycle. After a specific period (such as durability), the natural attributes of the items will change and then lose useable value and economic value; for the second category, the items have a short market life cycle. After a period of popularity in the market, the items lose the original economic value due to the changes in consumer preference, product upgrading and other reasons.[21]

The effect of deterioration is very important in many inventory systems. Deterioration is defined as decay or damage such that the item cannot be used for its original purpose. Most of the physical goods undergo decay or deterioration over time. Commodities such as fruits, vegetables, foodstuffs, etc., suffer from depletion by direct spoilage while kept in store. Highly volatile liquids such as gasoline, alcohol, turpentine, etc., undergo physical depletion over time through the process of evaporation. In the development economic production lot size models, usually researchers consider the deterioration rate, demand rate, unit cost, etc., as fixed, but all of them probably will have some little fluctuations for each cycle in real life situation. So in practical situations, if these quantities are treated as fuzzy variables then it will be more realistic.

The inventory problem of deteriorating items was first studied by Whitin , he studied fashion items deteriorating at the end of the storage period. Then Ghare and Schrader concluded in their study that the consumption of the deteriorating items was closely relative to a negative exponential function of time [37]. They proposed the deteriorating items inventory model as stated below:

$$\frac{dI(t)}{dt} + \theta I(t) = -D(t),$$

1.4 History:-

In 1915, the first inventory model was developed by F. Harris [5]. Later in 1965, first time the concept of fuzzy sets was introduced by Lotfi A Zadeh. Fuzzy set theory is an extension of classical set theory where elements have degrees of membership. The theory of fuzzy sets attracted the attention of many researches. In 1970, L. A. Zadeh and R. E. Bellman proposed a mathematical model on decision making in a fuzzy environment . In 1976, a fuzzy model on decision making in the presence of fuzzy variables was proposed by R. Jain [16]Later in 1978, D. Dubois and H. Prade defined some operations on fuzzy numbers. In 1983, H. J. Zimmerman made an attempt to use the fuzzy sets in operational research. Some researchers started to apply fuzzy set theory in inventory management problems. In 1982, J. Kacpryzk and P. Staniewski proposed a model on long-term inventory policy-making through fuzzy-decision making models and present a very interesting approach for aggregate inventory planning In the real world, fuzzy inventory model with both uncertain parameters and fuzzy variables has been discussed recently. Kacprzyk and Staniewski (1982). Park (1987) used fuzzy set concept to treat the inventory problem with fuzzy inventory cost under arithmetic operations of Extension Principle and proposed a model on fuzzy set theoretical interpretation of economic order quantity inventory problem. In 1996, M. Vujosevic, D. Petrovic and R. Petrovic developed an EOQ formula when inventory cost is a fuzzy number. In 1999, J. S. Yao and H. M. Lee proposed a fuzzy inventory with or without backorder for fuzzy order quantity with trapezoidal fuzzy number. In 1999, J. S. Yao and H. M. Lee developed an economic order quantity model in fuzzy sense for inventory without backorder model. In 2002, C. K. Kao and W. K. Hsu proposed a single-period inventory model with fuzzy demand. In 2002, C. H. Hsieh [5] proposed an approach for the optimization of fuzzy production inventory models. In 2003, J. S. Yao and J. Chiang introduced an inventory without back order with fuzzy total cost and fuzzy storing cost defuzzified by centroid and signed distance. They compared the optimal results obtained by both the defuzzification methods..

In 2003, Sujit De Kumar, P. K. Kundu and A. Goswami presented an economic production quantity inventory model involving fuzzy demand rate and fuzzy deterioration rate. In 2007, J. K. Syed and L. A. Aziz applied signed distance method to Fuzzy inventory model without shortages. In 2011, P. K. De and A. Rawat proposed a fuzzy inventory model without shortages using triangular fuzzy number. In 2012, C. K. Jaggi, S. Pareek, A. Sharma and Nidhi presented a fuzzy inventory model for deteriorating items with time-varying demand and shortages. In 2012,

Sumana Saha and Tripti Chakrabarti proposed a fuzzy EOQ model for time dependent deteriorating items and time dependent demand with shortages. Very recently, D. Dutta and Pavan Kumar published several papers in the area of fuzzy inventory with or with shortages. In 2012, presented a fuzzy inventory model without shortage using trapezoidal fuzzy number with sensitivity analysis. In 2013, the same authors D. Dutta and Pavan Kumar proposed an optimal policy for an inventory model without shortages considering fuzziness in demand, holding cost and ordering cost [5].

The classical inventory model (Narasimhan et al., 1995), EOQ, discussed in inventory control system assumed that demand and lead time are constant and known. Once we begin to move closer to reality, we must recognize that demands is never certain but that it occurs with some probability.

Yao and Lee introduced a backorder inventory model with fuzzy order quantity as triangular and trapezoidal fuzzy numbers and shortage cost as a crisp parameter. Gen et al. expressed their input data as fuzzy numbers, and then the interval mean value concept was introduced to solve the inventory problem. Chang et al. considered the backorder inventory problem with fuzzy backorder such that the backorder quantity is a triangular fuzzy number.

Chang [2] discussed the fuzzy production inventory model for fuzzify the product quantity as triangular fuzzy number. Lee and Yao proposed the inventory without backorder models in the fuzzy sense, where the order quantity is fuzzified as the triangular fuzzy number. Yao et al. assumed to be the order quantity and the total demand rate as triangular fuzzy numbers and obtained the fuzzy inventory model without shortages. Wu and Yao [2] fuzzified the order quantity and shortage quantity into triangular fuzzy numbers in an inventory model with backorder and they obtained the membership function of the fuzzy cost and its centroid. Yao and Chiang considered the total cost of inventory without backorder. They fuzzified the total demand and cost of storing one unit per day into triangular fuzzy numbers and defuzzify by the centroid and the signed distance methods. Dutta et al. developed a model in presence of fuzzy random variable demand where the optimum is achieved using a graded mean integration representation. Chang et al. developed the mixture inventory model involving variable lead-time with backorders and lost sales. First they fuzzify the random lead-time demand to be a fuzzy random variable and then fuzzify the total demand to be the triangular fuzzy number and derive the fuzzy total cost. By the centroid method of defuzzification, they estimate the total cost in the fuzzy sense. Wee et al. developed an optimal inventory model for items with imperfect quality and shortage backordering. Lin developed the inventory problem for a periodic review model with variable lead-time and fuzzified the expected demand shortage and backorder rate using signed distance method to defuzzify. Roy and Samanta discussed a fuzzy continuous review inventory model without backorder for deteriorating items in which the cycle time is taken as a symmetric fuzzy number. They used the signed distance method to fuzzify the total cost. Gani and Maheswari developed an EOQ model with imperfect quality items with shortages where defective rate, demand, holding cost, ordering cost and shortage cost are taken as triangular fuzzy numbers. Graded mean integration method is used for defuzzification of the total profit. Ameli et al. developed a new inventory model to determine ordering policy for imperfect items with fuzzy defective percentage under fuzzy discounting and inflationary conditions. They used the signed distance method of defuzzification to estimate the value of total profit. Nezhad et al.

developed a periodic review model and a continuous review inventory model with fuzzy setup cost, holding cost and shortage cost. Also they considered the lead-time demand and the lead-time plus one period's demand as random variables. They use two methods in the name of signed distance and possibility mean value to defuzzify. Uthayakumar and Valliathl developed an economic production model for Weibull deteriorating items over an infinite horizon under fuzzy environment and considered some cost component as triangular fuzzy numbers and using the signed distance method to defuzzify the cost function.[2]

Chapter-2
Definitions and Notations

2.1 Preliminaries (definitions and conditions relate to PFN)

2.2 Notations and Assumptions

2.1 Preliminaries (definitions and conditions relate to PFN)

[a] In order to treat fuzzy inventory model by using graded mean representation method and signed distance method to defuzzify, we need the following definitions:-

Definition 1: (By Pu and Liu [11]) A fuzzy set $\tilde{a}$ on R= $(-\infty, \infty)$ is called a fuzzy point if its membership function is

$$\mu_{\tilde{a}}(x) = \begin{cases} 1, x = a \\ 0, x \neq a \end{cases} \qquad (1)$$

Where the point a is called the support of fuzzy set $\tilde{a}$.

Definition 2-A fuzzy set $[a_\alpha, b_\alpha]$ where $0 \leq \alpha \leq 1$ and $a < b$ defined on R, is called a level of a fuzzy interval if its membership function is

$$\mu_{[a_\alpha, b_\alpha]}(x) = \begin{cases} \alpha, a \leq x \leq b \\ 0, otherwise \end{cases} \qquad (2)$$

Definition 3:- A fuzzy number $\tilde{A} = (a, b, c)$ where $a < b < c$ and defined on R, is called a triangular fuzzy number [2] if its membership function is

$$\mu_A = \begin{cases} \dfrac{x-a}{b-a}, a \leq x \leq b \\ \dfrac{c-x}{c-b}, b \leq x \leq c \\ 0, \ otherwise \end{cases} \qquad (3)$$

When $a = b = c$, we have fuzzy point $(c,c,c) = \tilde{c}$. The family of all triangular fuzzy numbers on R is denoted as

$$F_N = \{(a,b,c), a < b < c, \forall a,b,c \in R\}$$

The α-cut of $\tilde{A} = (a,b,c) \in F_N, 0 \leq \alpha \leq 1$ is $A(\alpha) = [A_L(\alpha), A_R(\alpha)]$

Where $A_L(\alpha) = a + (b-a)\alpha$ and $A_R(\alpha) = c - (c-b)\alpha$ are the left and right endpoints of $A(\alpha)$.

Definition 4: A trapezoidal fuzzy number $\tilde{A} = (a,b,c,d)$ is represented [4] with membership function $\mu_{\tilde{A}}$ as:

$$\mu_{\tilde{A}}(x) = \begin{cases} L(x) = \frac{x-a}{b-a}, & a \leq x \leq b \\ 1, & b \leq x \leq c \\ R(x) = \frac{d-x}{d-c}, & c \leq x \leq d \\ 0, & \text{otherwise} \end{cases}$$

(4)

The α-cut of $\tilde{A} = (a,b,c,d), 0 \leq \alpha \leq 1$ is $A(\alpha) = [A_L(\alpha), A_R(\alpha)]$

Where $A_L(\alpha) = a + (b-a)\alpha$ and $A_R(\alpha) = d - (d-c)\alpha$ are the left and right endpoints of $A(\alpha)$.

Definition 5: A pentagonal fuzzy number(PFN)[28] $\tilde{A} = (a,b,c,d,e)$ is represented with membership function $\mu_{\tilde{A}}$ as:

$$\mu_{\tilde{A}}(x) = \begin{cases} L_1(x) = \dfrac{x-a}{b-a}, a \leq x \leq b \\ L_2(x) = \dfrac{x-b}{c-b}, b \leq x \leq c \\ 1 \qquad\qquad , x = c \\ R_1(x) = \dfrac{d-x}{d-c}, c \leq x \leq d \\ R_2(x) = \dfrac{e-x}{e-d}, d \leq x \leq e \\ 0 \qquad\qquad , otherwise \end{cases} \qquad (5)$$

The α-cut of $\tilde{A} = (a,b,c,d,e), 0 \leq \alpha \leq 1$ is

$$A(\alpha) = [A_L(\alpha), A_R(\alpha)]$$

Where $A_{L_1}(\alpha) = a + (b-a)\alpha = L_1^{-1}(\alpha)$

$A_{L_2}(\alpha) = b + (c-b)\alpha = L_2^{-1}(\alpha)$

and

$A_{R_1}(\alpha) = d - (d-c)\alpha = R_1^{-1}(\alpha)$

$A_{R_2}(\alpha) = e - (e-d)\alpha = R_2^{-1}(\alpha)$

So

$$L^{-1}(\alpha) = \frac{L_1^{-1}(\alpha) + L_2^{-1}(\alpha)}{2} = \frac{a + (b-a)\alpha + b + (c-b)\alpha}{2}$$

$$= \frac{a + b + (b - a + c - b)\alpha}{2} = \frac{a + b + (c-a)\alpha}{2}$$

$$R^{-1}(\alpha) = \frac{R_1^{-1}(\alpha) + R_2^{-1}(\alpha)}{2} = \frac{d - (d-c)\alpha + e - (e-d)\alpha}{2}$$

$$= \frac{d + e - (d - c + e - d)\alpha}{2} = \frac{d + e - (e - c)\alpha}{2}$$

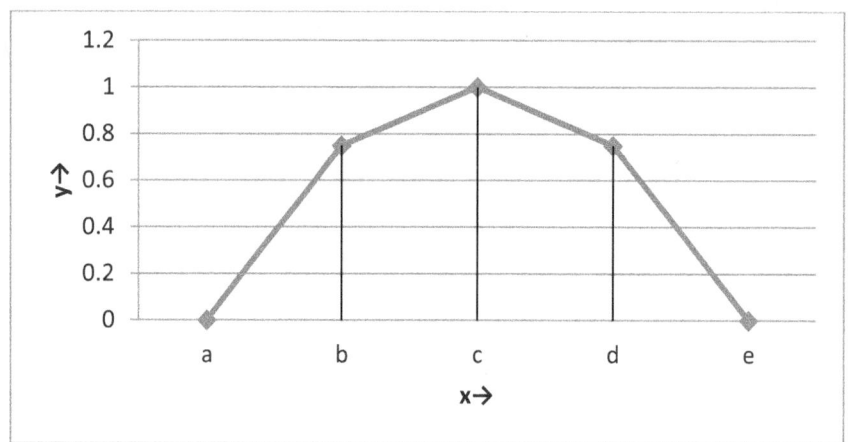

Fig (1): Graphical representation of Pentagonal Fuzzy Number (PFN)

Definition 6: If $\tilde{A} = (a, b, c, d, e)$ is a pentagonal fuzzy number then the graded mean integration representation of $\tilde{A}$ is defined as

$$P(\tilde{A}) = \frac{\int_0^{W_A} h \left(\frac{L^{-1}(h) + R^{-1}(h)}{2} \right) dh}{\int_0^{W_A} h \, dh}$$

With $\quad 0 \leq h \leq W_A \quad$ and $\quad 0 \leq W_A \leq 1$

$$P(\tilde{A}) = \frac{1}{2} \frac{\int_0^1 h \left[\frac{a + b + (c-a)h}{2} + \frac{d + e - (e-c)h}{2} \right] dh}{\int_0^1 h \, dh}$$

$$= \left[\left(\frac{a+b+d+e}{2}\right)\frac{h^2}{2} + \left(\frac{c-a-e+c}{2}\right)\frac{h^3}{3}\right]_0^1$$

$$= \left(\frac{a+b+d+e}{4}\right) + \left(\frac{c-a-e+c}{6}\right)$$

$$= \frac{a+3b+4c+3d+e}{12} \qquad (6)$$

Definition 7: If $\tilde{A} = (a, b, c, d, e)$ is a pentagonal fuzzy number then the signed distance method[5] of $\tilde{A}$ is defined as

$$d(\tilde{A}, \tilde{0}) = \int_0^1 d([A_L(\alpha)_\alpha, A_R(\alpha)_\alpha], \tilde{0})$$

$$= \frac{1}{2}\int_0^1 \left[\frac{a+b+(c-a)\alpha}{2} + \frac{d+e-(e-c)\alpha}{2}\right]d\alpha$$

$$= \frac{1}{4}\left[a+b+d+e+\frac{2c-a-e}{2}\right]$$

$$= \frac{1}{8}(a+2b+2c+2d+e) \qquad (7)$$

[b] Conditions on Pentagonal Fuzzy Number [PFN]:-

A Pentagonal Fuzzy Number $P(\tilde{A})$ should satisfy the following conditions [28];

1. $\mu_{\tilde{A}}(x)$ is a continuous function in the interval [0,1].
2. $\mu_{\tilde{A}}(x)$ is strictly increasing and continuous function on [a, b] and [b, c].
3. $\mu_{\tilde{A}}(x)$ is strictly decreasing and continuous function on [c, d] and [d, e].

[c] Arithmetical operations of PFN:-[28]

Suppose $\tilde{A} = (a_1, a_2, a_3, a_4, a_5)$ and $\tilde{B} = (b_1, b_2, b_3, b_4, b_5)$ are two pentagonal fuzzy numbers, then arithmetical operations are defined as:

1. $\tilde{A} \oplus \tilde{B} = (a_1 + b_1, a_2 + b_2, a_3 + b_3, a_4 + b_4, a_5 + b_5)$
2. $\tilde{A} \ominus \tilde{B} = (a_1 - b_1, a_2 - b_2, a_3 - b_3, a_4 - b_4, a_5 - b_5)$
3. $\tilde{A} \otimes \tilde{B} = (a_1 b_1, a_2 b_2, a_3 b_3, a_4 b_4, a_5 b_5)$
4. $\tilde{A} \oslash \tilde{B} = (\frac{a_1}{b_1}, \frac{a_2}{b_2}, \frac{a_3}{b_3}, \frac{a_4}{b_4}, \frac{a_5}{b_5})$
5. $\alpha \otimes \tilde{A} = \begin{cases} (\alpha a_1, \alpha a_2, \alpha a_3, \alpha a_4, \alpha a_5), \alpha \geq 0 \\ (\alpha a_5, \alpha a_4, \alpha a_3, \alpha a_2, \alpha a_1), \alpha < 0 \end{cases}$

2.2 Notations and Assumptions

The mathematical model in this paper is developed on the basis of the following assumptions and notations [2].

Notations:-

1. D (t) is the demand rate at any time t per unit time.
2. A is the ordering cost per order.
3. θ is the deterioration rate, $0 < \theta < 1$.
4. T is the length of the Cycle.
5. Q is the ordering Quantity per unit.

6. h is the holding cost per unit per unit time
7. S is the shortage Cost per unit time.
8. C is the unit Cost per unit time.
9. $K(t_1, T)$ is the total inventory cost per unit time.
10. $\widetilde{D}$ is the fuzzy demand.
11. $\widetilde{\theta}$ is the fuzzy deterioration rate.
12. $\widetilde{h}$ is the fuzzy holding cost per unit per unit time.
13. $\widetilde{S}$ is the fuzzy shortage Cost per unit time.
14. $\widetilde{C}$ is the fuzzy unit Cost per unit time.
15. $\widetilde{K}(t_1, T)$ is the total fuzzy inventory cost per unit time.
16. $K_{dG}(t_1, T)$ is the defuzzify value of $\widetilde{K}(t_1, T)$ by applying Graded mean representation method.
17. $K_{dS}(t_1, T)$ is the defuzzify value of $\widetilde{K}(t_1, T)$ by applying Signed Distance Method .

Assumptions:-
1. Demand $D(t) = a(1 + bt)$ is assumed to be an increasing function of time i.e. where a and b are positive constants and $a > 0, 0 < b < 1$.
2. Replenishment is instantaneous and lead time is zero.
3. Shortages are allowed and fully backlogged.

Chapter-3

Mathematical model and Numerical example

4.1 Mathematical Model

    4.1(a) Crisp Model

    4.1(b) Fuzzy Model

4.2 Numerical Example

4.1 MATHEMATICAL MODEL

Let Q be the total amount of inventory purchased or produced at the beginning of each period and after fulfilling backorders. Due to reasons of market demand and deterioration of the items, the inventory level gradually diminishes during the period $[0, t_1]$ and ultimately falls to zero at $t = t_1$. the period $[t_1, T]$ is the period of shortages, which are fully backlogged. Let $I[t]$ be the on-hand inventory level at any time t, which is governed by the following two differential equations [2]:

4.1 (a) **CRISP MODEL**:

$$\frac{dI(t)}{dt} + \theta I(t) = -D(t), \quad 0 \leq t \leq t_1 \tag{4.1}$$

with $I(0) = Q$, $I(t_1) = 0$

$$\frac{dI(t)}{dt} = -D(t), \quad t_1 \leq t \leq T \tag{4.2}$$

, with $I(t_1) = 0$

Solution of equation (4.1):-

$$\frac{dI(t)}{dt} + \theta I(t) = -D(t)$$

So $\quad$ I.F.$= e^{\int \theta dt} = e^{\theta t}$

$$I(t).e^{\theta t} = -\int a(1+bt).e^{\theta t}dt + C$$

$$I(t).e^{\theta t} = -\left[\frac{ae^{\theta t}}{\theta} + ab\left\{\frac{te^{\theta t}}{\theta} - \frac{e^{\theta t}}{\theta^2}\right\}\right] + C \tag{4.1a}$$

$$because\ I(0) = Q$$

$$\Rightarrow Q = -\left[\frac{a}{\theta} + ab\left\{-\frac{1}{\theta^2}\right\}\right] + C$$

$$\Rightarrow Q = -\frac{a}{\theta} + \frac{ab}{\theta^2} + C$$

$$C = Q + \frac{a}{\theta} - \frac{ab}{\theta^2}$$

Put the value of C in equation (4.1a), we get

$$I(t).e^{\theta t} = -\left[\frac{ae^{\theta t}}{\theta} + ab\left\{\frac{te^{\theta t}}{\theta} - \frac{e^{\theta t}}{\theta^2}\right\}\right] + Q + \frac{a}{\theta} - \frac{ab}{\theta^2}$$

$$I(t) = Qe^{-\theta t} + \left(\frac{a}{\theta} - \frac{ab}{\theta^2}\right)e^{-\theta t} - \frac{a}{\theta} - \frac{ab}{\theta}t + \frac{ab}{\theta^2}$$

$$I(t) = Qe^{-\theta t} + \left(\frac{a}{\theta} - \frac{ab}{\theta^2}\right)e^{-\theta t} + \frac{ab}{\theta^2} - \frac{a}{\theta}(1 + bt) \quad (4.3)$$

The solution of equation (4.2) is given by

$$\frac{dI(t)}{dt} = -D(t)$$

$$I(t) = -\int_{t_1}^{T} D(t)dt = -\int_{t_1}^{T} a(1 + bt)dt = -\left[at + \frac{abt^2}{2}\right]_{t_1}^{T}$$

$$= -\left[aT + \frac{abT^2}{2} - at_1 - \frac{abt_1^2}{2}\right]$$

$$I(t) = a(t_1 - T) + \frac{ab}{2}(t_1^2 - T^2) \quad (4.4)$$

By using $I(t_1) = 0$, put t= t_1 in equation (4.3), we get

$$I(t_1) = Qe^{-\theta t_1} + \left(\frac{a}{\theta} - \frac{ab}{\theta^2}\right)e^{-\theta t_1} + \frac{ab}{\theta^2} - \frac{a}{\theta}(1+bt_1) = 0$$

$$Q = \left\{\frac{a}{\theta}(1+bt_1) - \frac{ab}{\theta^2}\right\}e^{\theta t_1}$$
$$- \left(\frac{a}{\theta} - \frac{ab}{\theta^2}\right) \qquad (4.5)$$

Now (4.3) becomes

$$I(t) = \left[\frac{a}{\theta}(1+bt_1) - \frac{ab}{\theta^2}\right]e^{\theta t_1}\cdot e^{-\theta t} - \left(\frac{a}{\theta} - \frac{ab}{\theta^2}\right)e^{-\theta t}$$
$$+ \left(\frac{a}{\theta} - \frac{ab}{\theta^2}\right)e^{-\theta t} + \frac{ab}{\theta^2} - \frac{a}{\theta}(1+bt)$$

$$= \left[\frac{a}{\theta}(1+bt_1) - \frac{ab}{\theta^2}\right]e^{\theta(t_1-t)} + \frac{ab}{\theta^2} - \frac{a}{\theta}(1+bt)$$

$$= \left[\frac{a}{\theta}(1+bt_1) - \frac{ab}{\theta^2}\right]\left[1 + \theta(t_1-t) + \frac{\theta^2(t_1-t)^2}{!2} + \frac{\theta^3(t_1-t)^3}{!3}\right.$$
$$\left. + \cdots\right] + \frac{ab}{\theta^2} - \frac{a}{\theta}(1+bt)$$

$$= \frac{a}{\theta}(1+bt_1) - \frac{ab}{\theta^2} + a(1+bt_1)(t_1-t) - \frac{ab}{\theta}(t_1-t)$$

$$+ \frac{a\theta}{2}(1+bt_1)(t_1-t)^2 - \frac{ab}{2}(t_1-t)^2$$

$$+ \frac{a\theta^2}{6}(1+bt_1)(t_1-t)^3 - \frac{ab\theta}{6}(t_1-t)^3 + \cdots + \frac{ab}{\theta^2} - \frac{a}{\theta}(1+bt)$$

By neglecting higher term of θ

$$I(t) = \frac{a}{\theta}(1 + bt_1 - 1 - bt) + a(t_1 - t) + abt_1(t_1 - t) - \frac{ab}{\theta}(t_1 - t)$$

$$+ \frac{a\theta}{2}(t_1 - t)^2 + \frac{ab\theta}{2}t_1(t_1 - t)^2 - \frac{ab}{2}(t_1 - t)^2 - \frac{ab\theta}{6}(t_1 - t)^3$$

$$I(t) = a\left\{(t_1 - t) + \frac{\theta}{2}(t_1 - t)^2\right\}$$

$$+ ab\left\{t_1(t_1 - t) - \frac{(t_1 - t)^2}{2} + \frac{\theta}{2}t_1(t_1 - t)^2 - \frac{\theta}{6}(t_1 - t)^3\right\} \quad (4.6)$$

Total average no. of holding units (I_H) during period [0, T] is given by

$$I_H = \int_0^{t_1} I(t)\, dt$$

$$= \int_0^{t_1} \left[a\left\{(t_1 - t) + \frac{\theta}{2}(t_1 - t)^2\right\}\right.$$

$$\left. + ab\left\{t_1(t_1 - t) - \frac{(t_1 - t)^2}{2} + \frac{\theta}{2}t_1(t_1 - t)^2 - \frac{\theta}{6}(t_1 - t)^3\right\}\right] dt$$

$$= \left[a\left\{-\frac{(t_1 - t)^2}{2} - \frac{\theta(t_1 - t)^3}{2 \cdot 3}\right\} + ab\left\{-t_1\frac{(t_1 - t)^2}{2} + \frac{(t_1 - t)^3}{6} - \frac{\theta}{2}t_1\frac{(t_1 - t)^3}{3} + \frac{\theta(t_1 - t)^4}{6 \cdot 4}\right\}\right]_0^{t_1}$$

$$= a\left\{\frac{t_1^2}{2} + \frac{\theta}{6}t_1^3\right\} + ab\left\{\frac{t_1^3}{2} - \frac{t_1^3}{6} + \frac{\theta}{6}t_1^4 - \frac{\theta}{24}t_1^4\right\}$$

$$= a\left\{\frac{t_1^2}{2} + \frac{\theta}{6}t_1^3\right\} + ab\left\{\frac{t_1^3}{3} + \frac{\theta}{8}t_1^4\right\} \quad (4.7)$$

Total no. of deteriorated units (I_D) during period [0, T] is given by

$$I_D = Q - \text{Total demand} = Q - \int_0^{t_1} a(1 + bt)\, dt$$

$$= \left\{\frac{a}{\theta}(1+bt_1) - \frac{ab}{\theta^2}\right\}e^{\theta t_1} - \left(\frac{a}{\theta} - \frac{ab}{\theta^2}\right) - \left(at + \frac{abt^2}{2}\right)\Big|_0^{t_1}$$

$$= \left\{\frac{a}{\theta}(1+bt_1) - \frac{ab}{\theta^2}\right\}\left\{1 + \frac{\theta t_1}{1!} + \frac{(\theta t_1)^2}{2!} + \frac{(\theta t_1)^3}{3!} + \cdots\right\}$$

$$- \left(\frac{a}{\theta} - \frac{ab}{\theta^2}\right) - \left(at_1 + \frac{abt_1^2}{2}\right)$$

$$= \frac{a}{\theta}(1+bt_1) - \frac{ab}{\theta^2} + at_1(1+bt_1) - \frac{ab}{\theta}t_1 + \frac{a\theta t_1^2}{2}(1+bt_1)$$

$$- \frac{abt_1^2}{2} + \frac{a\theta^2}{6}(1+bt_1) - \frac{ab\theta t_1^3}{6} + \cdots - \frac{a}{\theta} + \frac{ab}{\theta^2} - at_1 - \frac{abt_1^2}{2}$$

Neglecting higher term of θ

$$= \frac{a}{\theta} + \frac{abt_1}{\theta} + at_1 + abt_1^2 - \frac{ab}{\theta}t_1 + \frac{a\theta t_1^2}{2}(1+bt_1) - \frac{abt_1^2}{2}$$

$$- \frac{ab\theta t_1^3}{6} - \frac{a}{\theta} - at_1 - \frac{abt_1^2}{2}$$

$$= \frac{a\theta t_1^2}{2} + \frac{ab\theta}{2}t_1^3 - \frac{ab\theta}{6}t_1^3$$

$$I_D$$

$$= \frac{a\theta t_1^2}{2}$$

$$+ \frac{ab\theta}{3}t_1^3 \tag{4.8}$$

Total average no. of shortage units (I_S) during period [0, T] is given by

$$I_S = -\int_{t_1}^{T} I(t)\, dt$$

$$= -\int_{t_1}^{T}\left[a\left\{(t_1-t)+\frac{\theta}{2}(t_1-t)^2\right\}+ab\left\{t_1(t_1-t)-\frac{(t_1-t)^2}{2}+\frac{\theta}{2}t_1(t_1-t)^2-\frac{\theta}{6}(t_1-t)^3\right\}\right]dt$$

$$= -\left[a\left\{-\frac{(t_1-t)^2}{2}-\frac{\theta(t_1-t)^3}{2\cdot 3}\right\}+ab\left\{-t_1\frac{(t_1-t)^2}{2}+\frac{(t_1-t)^3}{6}-\frac{\theta}{2}t_1\frac{(t_1-t)^3}{3}+\frac{\theta(t_1-t)^4}{6\cdot 4}\right\}\right]_{t_1}^{T}$$

$$= -\left[a\left\{-\frac{(t_1-T)^2}{2}-\frac{\theta(t_1-T)^3}{2\cdot 3}\right\}+ab\left\{-t_1\frac{(t_1-T)^2}{2}+\frac{(t_1-T)^3}{6}-\frac{\theta}{2}t_1\frac{(t_1-T)^3}{3}+\frac{\theta(t_1-T)^4}{6\cdot 4}\right\}\right]$$

$$= \frac{a}{2}(t_1-T)^2+\frac{a\theta}{2}(t_1-T)^3-ab\frac{(t_1-T)^2}{2}\left[-t_1+\frac{(t_1-T)}{3}+\frac{\theta}{2}t_1\frac{(t_1-T)^3}{3}-\frac{\theta(t_1-T)^4}{6\cdot 4}\right]$$

$$= \frac{a}{2}(t_1-T)^2-\frac{ab}{2}(t_1^2+T^2-2t_1T)\left(\frac{-3t_1+t_1-T}{3}\right)$$

$$= \frac{a}{2}(t_1-T)^2-\frac{ab}{2}(t_1^2+T^2-2t_1T)\left(\frac{-2t_1-T}{3}\right)$$

$$= \frac{a}{2}(t_1-T)^2+\frac{ab}{2}\left(\frac{2t_1^3+t_1^2+2t_1T^2+T^3-4t_1^2T-2t_1T^2}{3}\right)$$

$$= \frac{a}{2}(t_1-T)^2-\frac{ab}{2}\left\{t_1^2T-\frac{T^3}{3}-\frac{2}{3}t_1^3\right\} \qquad (4.9)$$

Total cost of the system per unit time is given by

$$K(t_1,T)=\frac{1}{T}[A+hI_H+CI_D+SI_S]$$

$$K(t_1,T)=\frac{1}{T}\left[A+ha\left\{\frac{t_1^2}{2}+\frac{\theta}{6}t_1^3\right\}+hab\left\{\frac{t_1^3}{3}+\frac{\theta}{8}t_1^4\right\}\right.$$

$$+C\left\{\frac{a\theta t_1^2}{2}+\frac{ab\theta}{3}t_1^3\right\}+S\left\{\frac{a}{2}(t_1-T)^2-\frac{ab}{2}\left(t_1^2T-\frac{T^3}{3}-\frac{2}{3}t_1^3\right)\right\}$$

$$(4.10)$$

4.1(b) Fuzzy Model

Due to uncertainly in the environment it is not easy to define all the parameters precisely, accordingly we assume some of these parameters viz. $\tilde{a}, \tilde{b}, \tilde{C}, \tilde{S}, \tilde{\theta}, \tilde{h}$ may change within some limits.

Let

$$\tilde{a} = (a_1, a_2, a_3, a_4, a_5)$$

$$\tilde{b} = (b_1, b_2, b_3, b_4, b_5)$$
$$\tilde{C} = (C_1, C_2, C_3, C_4, C_5)$$

$$\tilde{S} = (S_1, S_2, S_3, S_4, S_5)$$

$$\tilde{\theta} = (\theta_1, \theta_2, \theta_3, \theta_4, \theta_5)$$

$\tilde{h} = (h_1, h_2, h_3, h_4, h_5)$ are the pentagonal fuzzy numbers.

Total cost of the system per unit time in fuzzy sense is given by

$$\tilde{K}(t_1, T) = \frac{1}{T}\left[A + \tilde{h}\tilde{a}\left\{\frac{t_1^2}{2} + \frac{\tilde{\theta}}{6}t_1^3\right\} + \tilde{h}\tilde{a}\tilde{b}\left\{\frac{t_1^3}{3} + \frac{\tilde{\theta}}{8}t_1^4\right\} + \tilde{C}\left\{\frac{\tilde{a}\tilde{\theta}t_1^2}{2} + \frac{\tilde{a}\tilde{b}\tilde{\theta}}{3}t_1^3\right\} \right.$$
$$\left. + \tilde{S}\left\{\frac{\tilde{a}}{2}(t_1 - T)^2 - \frac{\tilde{a}\tilde{b}}{2}\left(t_1^2 T - \frac{T^3}{3} - \frac{2}{3}t_1^3\right)\right\}\right] \quad (4.11)$$

We defuzzify the fuzzy total cost $\tilde{K}(t_1, T)$ by graded mean representation method and signed distance methods.

(i) By Graded Mean Representation Method, Total Cost is given by.

$$K_{dG}(t_1, T) = \frac{1}{12}\left[K_{dG_1}(t_1, T), K_{dG_2}(t_1, T), K_{dG_3}(t_1, T), K_{dG_4}(t_1, T), K_{dG_5}(t_1, T)\right]$$

Where

$$K_{dG_1}(t_1, T) = \frac{1}{T}\left[A + h_1 a_1 \left\{\frac{t_1^2}{2} + \frac{\theta_1}{6}t_1^3\right\} + h_1 a_1 b_1 \left\{\frac{t_1^3}{3} + \frac{\theta_1}{8}t_1^4\right\}\right.$$

$$+ C_1 \left\{\frac{a_1 \theta_1 t_1^2}{2} + \frac{a_1 b_1 \theta_1}{3}t_1^3\right\}$$

$$\left. + S_1 \left\{\frac{a_1}{2}(t_1 - T)^2 - \frac{a_1 b_1}{2}\left(t_1^2 T - \frac{T^3}{3} - \frac{2}{3}t_1^3\right)\right\}\right]$$

$$K_{dG_2}(t_1, T) = \frac{1}{T}\left[A + h_2 a_2 \left\{\frac{t_1^2}{2} + \frac{\theta_2}{6}t_1^3\right\} + h_2 a_2 b_2 \left\{\frac{t_1^3}{3} + \frac{\theta_2}{8}t_1^4\right\}\right.$$

$$+ C_2 \left\{\frac{a_2 \theta_2 t_1^2}{2} + \frac{a_2 b_2 \theta_2}{3}t_1^3\right\}$$

$$\left. + S_2 \left\{\frac{a_2}{2}(t_1 - T)^2 - \frac{a_2 b_2}{2}\left(t_1^2 T - \frac{T^3}{3} - \frac{2}{3}t_1^3\right)\right\}\right]$$

$$K_{dG_3}(t_1, T) = \frac{1}{T}\left[A + h_3 a_3 \left\{\frac{t_1^2}{2} + \frac{\theta_3}{6}t_1^3\right\} + h_3 a_3 b_3 \left\{\frac{t_1^3}{3} + \frac{\theta_3}{8}t_1^4\right\}\right.$$

$$+ C_3 \left\{\frac{a_3 \theta_3 t_1^2}{2} + \frac{a_3 b_3 \theta_3}{3}t_1^3\right\}$$

$$\left. + S_3 \left\{\frac{a_3}{2}(t_1 - T)^2 - \frac{a_3 b_3}{2}\left(t_1^2 T - \frac{T^3}{3} - \frac{2}{3}t_1^3\right)\right\}\right]$$

$$K_{dG_4}(t_1, T) = \frac{1}{T}\left[A + h_4 a_4 \left\{\frac{t_1^2}{2} + \frac{\theta_4}{6}t_1^3\right\} + h_4 a_4 b_4 \left\{\frac{t_1^3}{3} + \frac{\theta_4}{8}t_1^4\right\}\right.$$

$$+ C_4 \left\{\frac{a_4 \theta_4 t_1^2}{2} + \frac{a_4 b_4 \theta_4}{3}t_1^3\right\}$$

$$\left. + S_4 \left\{\frac{a_4}{2}(t_1 - T)^2 - \frac{a_4 b_4}{2}\left(t_1^2 T - \frac{T^3}{3} - \frac{2}{3}t_1^3\right)\right\}\right]$$

$$K_{dG_5}(t_1,T) = \frac{1}{T}\left[A + h_5 a_5 \left\{\frac{t_1^2}{2} + \frac{\theta_5}{6}t_1^3\right\} + h_5 a_5 b_5 \left\{\frac{t_1^3}{3} + \frac{\theta_5}{8}t_1^4\right\}\right.$$

$$+ C_5 \left\{\frac{a_5 \theta_5 t_1^2}{2} + \frac{a_5 b_5 \theta_5}{3}t_1^3\right\}$$

$$+ S_5 \left\{\frac{a_5}{2}(t_1 - T)^2\right.$$

$$\left.\left.- \frac{a_5 b_5}{2}\left(t_1^2 T - \frac{T^3}{3} - \frac{2}{3}t_1^3\right)\right\}\right] \quad (4.12)$$

$$K_{dG}(t_1,T) = \frac{1}{12}\left[K_{dG_1}(t_1,T) + 3K_{dG_2}(t_1,T) + 4K_{dG_3}(t_1,T)\right.$$

$$\left. + 3K_{dG_4}(t_1,T) + K_{dG_5}(t_1,T)\right]$$

To minimize total cost function per unit time $K_{dG}(t_1,T)$, the optimal value of t_1 and T can be obtained by solving the following equations:

$$\frac{\partial K_{dG}(t_1,T)}{\partial t_1} = 0$$

and

$$\frac{\partial K_{dG}(t_1,T)}{\partial T} = 0 \quad (4.13)$$

Equation (4.13) is equivalent to

$$\frac{1}{12T}\left[h_1 a_1\left\{t_1+\frac{\theta_1}{2}t_1^2\right\}+h_1 a_1 b_1\left\{t_1^2+\frac{\theta_1}{2}t_1^3\right\}\right.$$

$$+C_1\{a_1\theta_1 t_1+a_1 b_1\theta_1 t_1^2\}+S_1\{a_1(t_1-T)-a_1 b_1(t_1 T-t_1^2)\}$$

$$+3\left\{h_2 a_2\left\{t_1+\frac{\theta_2}{2}t_1^2\right\}+h_2 a_2 b_2\left\{t_1^2+\frac{\theta_2}{2}t_1^3\right\}\right.$$

$$\left.+C_2\{a_2\theta_2 t_1+a_2 b_2\theta_2 t_1^2\}+S_2\{a_2(t_1-T)-a_2 b_2(t_1 T-t_1^2)\}\right\}$$

$$+4\left\{h_3 a_3\left\{t_1+\frac{\theta_3}{2}t_1^2\right\}+h_3 a_3 b_3\left\{t_1^2+\frac{\theta_3}{2}t_1^3\right\}\right.$$

$$\left.+C_3\{a_3\theta_3 t_1+a_3 b_3\theta_3 t_1^2\}+S_3\{a_3(t_1-T)-a_3 b_3(t_1 T-t_1^2)\}\right\}$$

$$+3\left\{h_4 a_4\left\{t_1+\frac{\theta_4}{2}t_1^2\right\}+h_4 a_4 b_4\left\{t_1^2+\frac{\theta_4}{2}t_1^3\right\}\right.$$

$$\left.+C_4\{a_4\theta_4 t_1+a_4 b_4\theta_4 t_1^2\}+S_4\{a_4(t_1-T)-a_4 b_4(t_1 T-t_1^2)\}\right\}$$

$$+h_5 a_5\left\{t_1+\frac{\theta_5}{2}t_1^2\right\}+h_5 a_5 b_5\left\{t_1^2+\frac{\theta_5}{2}t_1^3\right\}$$

$$\left.+C_5\{a_5\theta_5 t_1+a_5 b_5\theta_5 t_1^2\}+S_5\{a_5(t_1-T)-a_5 b_5(t_1 T-t_1^2)\}\right]$$

$$=0 \tag{4.14}$$

$$\text{And } \left[\frac{1}{12T} \left\{ S_1 \left\{ -a_1(t_1 - T) - \frac{a_1 b_1}{2}(t_1^2 - T^2) \right\} \right. \right.$$

$$+ 3S_2 \left\{ -a_2(t_1 - T) - \frac{a_2 b_2}{2}(t_1^2 - T^2) \right\}$$

$$+ 4S_3 \left\{ -a_3(t_1 - T) - \frac{a_3 b_3}{2}(t_1^2 - T^2) \right\}$$

$$+ 3S_4 \left\{ -a_4(t_1 - T) - \frac{a_4 b_4}{2}(t_1^2 - T^2) \right\}$$

$$+ S_5 \left\{ -a_5(t_1 - T) - \frac{a_5 b_5}{2}(t_1^2 - T^2) \right\} \right\}$$

$$- \frac{1}{12T^2} \left\{ 12A + h_1 a_1 \left\{ \frac{t_1^2}{2} + \frac{\theta_1}{6} t_1^3 \right\} \right.$$

$$+ h_1 a_1 b_1 \left\{ \frac{t_1^3}{3} + \frac{\theta_1}{8} t_1^4 \right\}$$

$$+ C_1 \left\{ \frac{a_1 \theta_1 t_1^2}{2} + \frac{a_1 b_1 \theta_1}{3} t_1^3 \right\}$$

$$+ S_1 \left\{ \frac{a_1}{2}(t_1 - T)^2 \right.$$

$$\left. - \frac{a_1 b_1}{2} \left(t_1^2 T - \frac{T^3}{3} - \frac{2}{3} t_1^3 \right) \right\}$$

$$+ 3 \left\{ h_2 a_2 \left\{ \frac{t_1^2}{2} + \frac{\theta_2}{6} t_1^3 \right\} \right.$$

$$+ h_2 a_2 b_2 \left\{ \frac{t_1^3}{3} + \frac{\theta_2}{8} t_1^4 \right\}$$

$$+ C_2 \left\{ \frac{a_2 \theta_2 t_1^2}{2} + \frac{a_2 b_2 \theta_2}{3} t_1^3 \right\}$$

$$+ S_2 \left\{ \frac{a_2}{2}(t_1 - T)^2 \right.$$

$$-\frac{a_2 b_2}{2}\left(t_1^2 T - \frac{T^3}{3} - \frac{2}{3}t_1^3\right)\bigg\}\bigg\}$$

$$+ 4\left\{h_3 a_3 \left\{\frac{t_1^2}{2} + \frac{\theta_3}{6}t_1^3\right\}\right.$$

$$+ h_3 a_3 b_3 \left\{\frac{t_1^3}{3} + \frac{\theta_3}{8}t_1^4\right\}$$

$$+ C_3 \left\{\frac{a_3 \theta_3 t_1^2}{2} + \frac{a_3 b_3 \theta_3}{3}t_1^3\right\}$$

$$+ S_3 \left\{\frac{a_3}{2}(t_1 - T)^2\right.$$

$$-\frac{a_3 b_3}{2}\left(t_1^2 T - \frac{T^3}{3} - \frac{2}{3}t_1^3\right)\bigg\}\bigg\}$$

$$+ 3\left\{h_4 a_4 \left\{\frac{t_1^2}{2} + \frac{\theta_4}{6}t_1^3\right\}\right.$$

$$+ h_4 a_4 b_4 \left\{\frac{t_1^3}{3} + \frac{\theta_4}{8}t_1^4\right\}$$

$$+ C_4 \left\{\frac{a_4 \theta_4 t_1^2}{2} + \frac{a_4 b_4 \theta_4}{3}t_1^3\right\}$$

$$+ S_4 \left\{\frac{a_4}{2}(t_1 - T)^2\right.$$

$$-\frac{a_4 b_4}{2}\left(t_1^2 T - \frac{T^3}{3} - \frac{2}{3}t_1^3\right)\bigg\}\bigg\}$$

$$+ h_5 a_5 \left\{\frac{t_1^2}{2} + \frac{\theta_5}{6}t_1^3\right\} + h_5 a_5 b_5 \left\{\frac{t_1^3}{3} + \frac{\theta_5}{8}t_1^4\right\}$$

$$+ C_5 \left\{\frac{a_5 \theta_5 t_1^2}{2} + \frac{a_5 b_5 \theta_5}{3}t_1^3\right\}$$

$$+ S_5 \left\{\frac{a_5}{2}(t_1 - T)^2\right.$$

$$-\frac{a_5 b_5}{2}\left(t_1^2 T - \frac{T^3}{3} - \frac{2}{3}t_1^3\right)\Bigg\}\Bigg]$$

$$= 0 \qquad (4.15)$$

Further, for the total cost function $K_{dG}(t_1, T)$ to be convex, the following conditions must be satisfied

$$\frac{\partial^2 K_{dG}(t_1, T)}{\partial t_1^2} > 0, \frac{\partial^2 K_{dG}(t_1, T)}{\partial T^2} > 0 \qquad (4.16)$$

And

$$\left(\frac{\partial^2 K_{dG}(t_1, T)}{\partial t_1^2}\right)\left(\frac{\partial^2 K_{dG}(t_1, T)}{\partial T^2}\right) - \left(\frac{\partial^2 K_{dG}(t_1, T)}{\partial t_1 \partial T}\right)$$
$$> 0 \qquad (4.17)$$

The second derivatives of the total cost function $K_{dG}(t_1, T)$ are complicated and it is very difficult to prove the convexity mathematically. However, with the help of graph, we can easily demonstrate convexity of total fuzzy cost function.

(ii) We defuzzify the fuzzy total cost $\widetilde{K}(t_1, T)$ by signed distance method.

By signed distance method, Total Cost is given by.

$$K_{dS}(t_1, T)$$
$$= \frac{1}{8}\left[K_{dS_1}(t_1, T), K_{dS_2}(t_1, T), K_{dS_3}(t_1, T), K_{dS_4}(t_1, T), K_{dS_5}(t_1, T)\right]$$

Where

$$K_{dS_1}(t_1,T) = \frac{1}{T}\left[A + h_1 a_1\left\{\frac{t_1^2}{2} + \frac{\theta_1}{6}t_1^3\right\} + h_1 a_1 b_1\left\{\frac{t_1^3}{3} + \frac{\theta_1}{8}t_1^4\right\}\right.$$

$$+ C_1\left\{\frac{a_1\theta_1 t_1^2}{2} + \frac{a_1 b_1 \theta_1}{3}t_1^3\right\}$$

$$\left.+ S_1\left\{\frac{a_1}{2}(t_1-T)^2 - \frac{a_1 b_1}{2}\left(t_1^2 T - \frac{T^3}{3} - \frac{2}{3}t_1^3\right)\right\}\right]$$

$$K_{dS_2}(t_1,T) = \frac{1}{T}\left[A + h_2 a_2\left\{\frac{t_1^2}{2} + \frac{\theta_2}{6}t_1^3\right\} + h_2 a_2 b_2\left\{\frac{t_1^3}{3} + \frac{\theta_2}{8}t_1^4\right\}\right.$$

$$+ C_2\left\{\frac{a_2\theta_2 t_1^2}{2} + \frac{a_2 b_2 \theta_2}{3}t_1^3\right\}$$

$$\left.+ S_2\left\{\frac{a_2}{2}(t_1-T)^2 - \frac{a_2 b_2}{2}\left(t_1^2 T - \frac{T^3}{3} - \frac{2}{3}t_1^3\right)\right\}\right]$$

$$K_{dS_3}(t_1,T) = \frac{1}{T}\left[A + h_3 a_3\left\{\frac{t_1^2}{2} + \frac{\theta_3}{6}t_1^3\right\} + h_3 a_3 b_3\left\{\frac{t_1^3}{3} + \frac{\theta_3}{8}t_1^4\right\}\right.$$

$$+ C_3\left\{\frac{a_3\theta_3 t_1^2}{2} + \frac{a_3 b_3 \theta_3}{3}t_1^3\right\}$$

$$\left.+ S_3\left\{\frac{a_3}{2}(t_1-T)^2 - \frac{a_3 b_3}{2}\left(t_1^2 T - \frac{T^3}{3} - \frac{2}{3}t_1^3\right)\right\}\right]$$

$$K_{dS_4}(t_1,T) = \frac{1}{T}\left[A + h_4 a_4\left\{\frac{t_1^2}{2} + \frac{\theta_4}{6}t_1^3\right\} + h_4 a_4 b_4\left\{\frac{t_1^3}{3} + \frac{\theta_4}{8}t_1^4\right\}\right.$$

$$+ C_4\left\{\frac{a_4\theta_4 t_1^2}{2} + \frac{a_4 b_4 \theta_4}{3}t_1^3\right\}$$

$$\left.+ S_4\left\{\frac{a_4}{2}(t_1-T)^2 - \frac{a_4 b_4}{2}\left(t_1^2 T - \frac{T^3}{3} - \frac{2}{3}t_1^3\right)\right\}\right]$$

$$K_{dS_5}(t_1,T) = \frac{1}{T}\left[A + h_5 a_5 \left\{\frac{t_1^2}{2} + \frac{\theta_5}{6}t_1^3\right\} + h_5 a_5 b_5 \left\{\frac{t_1^3}{3} + \frac{\theta_5}{8}t_1^4\right\}\right.$$

$$+ C_5 \left\{\frac{a_5 \theta_5 t_1^2}{2} + \frac{a_5 b_5 \theta_5}{3}t_1^3\right\}$$

$$+ S_5 \left\{\frac{a_5}{2}(t_1 - T)^2\right.$$

$$\left.\left.- \frac{a_5 b_5}{2}\left(t_1^2 T - \frac{T^3}{3} - \frac{2}{3}t_1^3\right)\right\}\right] \qquad (4.12)$$

$$K_{dS}(t_1,T) = \frac{1}{8}\left[K_{dS_1}(t_1,T) + 2K_{dS_2}(t_1,T) + 2K_{dS_3}(t_1,T)\right.$$

$$\left. + 2K_{dS_4}(t_1,T) + K_{dS_5}(t_1,T)\right]$$

To minimize total cost function per unit time $K_{dS}(t_1,T)$, the optimal value of t_1 and T can be obtained by solving the following equations:

$$\frac{\partial K_{dS}(t_1,T)}{\partial t_1} = 0$$

and

$$\frac{\partial K_{dS}(t_1,T)}{\partial T} = 0 \qquad (4.13)$$

Equation (4.13) is equivalent to

$$\frac{1}{8T}\left[h_1 a_1\left\{t_1 + \frac{\theta_1}{2}t_1{}^2\right\} + h_1 a_1 b_1\left\{t_1{}^2 + \frac{\theta_1}{2}t_1{}^3\right\}\right.$$

$$+ C_1\{a_1\theta_1 t_1 + a_1 b_1 \theta_1 t_1{}^2\} + S_1\{a_1(t_1 - T) - a_1 b_1(t_1 T - t_1{}^2)\}$$

$$+ 2\left\{h_2 a_2\left\{t_1 + \frac{\theta_2}{2}t_1{}^2\right\} + h_2 a_2 b_2\left\{t_1{}^2 + \frac{\theta_2}{2}t_1{}^3\right\}\right.$$

$$+ C_2\{a_2\theta_2 t_1 + a_2 b_2 \theta_2 t_1{}^2\} + S_2\{a_2(t_1 - T) - a_2 b_2(t_1 T - t_1{}^2)\}\Big\}$$

$$+ 2\left\{h_3 a_3\left\{t_1 + \frac{\theta_3}{2}t_1{}^2\right\} + h_3 a_3 b_3\left\{t_1{}^2 + \frac{\theta_3}{2}t_1{}^3\right\}\right.$$

$$+ C_3\{a_3\theta_3 t_1 + a_3 b_3 \theta_3 t_1{}^2\} + S_3\{a_3(t_1 - T) - a_3 b_3(t_1 T - t_1{}^2)\}\Big\}$$

$$+ 2\left\{h_4 a_4\left\{t_1 + \frac{\theta_4}{2}t_1{}^2\right\} + h_4 a_4 b_4\left\{t_1{}^2 + \frac{\theta_4}{2}t_1{}^3\right\}\right.$$

$$+ C_4\{a_4\theta_4 t_1 + a_4 b_4 \theta_4 t_1{}^2\} + S_4\{a_4(t_1 - T) - a_4 b_4(t_1 T - t_1{}^2)\}\Big\}$$

$$+ h_5 a_5\left\{t_1 + \frac{\theta_5}{2}t_1{}^2\right\} + h_5 a_5 b_5\left\{t_1{}^2 + \frac{\theta_5}{2}t_1{}^3\right\}$$

$$+ C_5\{a_5\theta_5 t_1 + a_5 b_5 \theta_5 t_1{}^2\} + S_5\{a_5(t_1 - T) - a_5 b_5(t_1 T - t_1{}^2)\}\Bigg]$$

$$= 0 \tag{4.14}$$

And $\left[\frac{1}{8T}\left\{S_1\left\{-a_1(t_1-T)-\frac{a_1b_1}{2}(t_1{}^2-T^2)\right\}\right.\right.$

$$+2S_2\left\{-a_2(t_1-T)-\frac{a_2b_2}{2}(t_1{}^2-T^2)\right\}$$

$$+2S_3\left\{-a_3(t_1-T)-\frac{a_3b_3}{2}(t_1{}^2-T^2)\right\}$$

$$+2S_4\left\{-a_4(t_1-T)-\frac{a_4b_4}{2}(t_1{}^2-T^2)\right\}$$

$$+S_5\left\{-a_5(t_1-T)-\frac{a_5b_5}{2}(t_1{}^2-T^2)\right\}\right\}$$

$$-\frac{1}{8T^2}\left\{8A+h_1a_1\left\{\frac{t_1{}^2}{2}+\frac{\theta_1}{6}t_1{}^3\right\}\right.$$

$$+h_1a_1b_1\left\{\frac{t_1{}^3}{3}+\frac{\theta_1}{8}t_1{}^4\right\}$$

$$+C_1\left\{\frac{a_1\theta_1t_1{}^2}{2}+\frac{a_1b_1\theta_1}{3}t_1{}^3\right\}$$

$$+S_1\left\{\frac{a_1}{2}(t_1-T)^2\right.$$

$$\left.\left.-\frac{a_1b_1}{2}\left(t_1{}^2T-\frac{T^3}{3}-\frac{2}{3}t_1{}^3\right)\right)\right\}\right\}$$

$$+2\left\{h_2a_2\left\{\frac{t_1{}^2}{2}+\frac{\theta_2}{6}t_1{}^3\right\}\right.$$

$$+h_2a_2b_2\left\{\frac{t_1{}^3}{3}+\frac{\theta_2}{8}t_1{}^4\right\}$$

$$+C_2\left\{\frac{a_2\theta_2t_1{}^2}{2}+\frac{a_2b_2\theta_2}{3}t_1{}^3\right\}$$

$$+S_2\left\{\frac{a_2}{2}(t_1-T)^2\right.$$

$$-\frac{a_2 b_2}{2}\left(t_1{}^2 T - \frac{T^3}{3} - \frac{2}{3}t_1{}^3\right)\Big\}\Big\}$$

$$+ 2\left\{h_3 a_3 \left\{\frac{t_1{}^2}{2} + \frac{\theta_3}{6}t_1{}^3\right\}\right.$$

$$+ h_3 a_3 b_3 \left\{\frac{t_1{}^3}{3} + \frac{\theta_3}{8}t_1{}^4\right\}$$

$$+ C_3 \left\{\frac{a_3 \theta_3 t_1{}^2}{2} + \frac{a_3 b_3 \theta_3}{3}t_1{}^3\right\}$$

$$+ S_3 \left\{\frac{a_3}{2}(t_1 - T)^2\right.$$

$$-\frac{a_3 b_3}{2}\left(t_1{}^2 T - \frac{T^3}{3} - \frac{2}{3}t_1{}^3\right)\Big\}\Big\}$$

$$+ 2\left\{h_4 a_4 \left\{\frac{t_1{}^2}{2} + \frac{\theta_4}{6}t_1{}^3\right\}\right.$$

$$+ h_4 a_4 b_4 \left\{\frac{t_1{}^3}{3} + \frac{\theta_4}{8}t_1{}^4\right\}$$

$$+ C_4 \left\{\frac{a_4 \theta_4 t_1{}^2}{2} + \frac{a_4 b_4 \theta_4}{3}t_1{}^3\right\}$$

$$+ S_4 \left\{\frac{a_4}{2}(t_1 - T)^2\right.$$

$$-\frac{a_4 b_4}{2}\left(t_1{}^2 T - \frac{T^3}{3} - \frac{2}{3}t_1{}^3\right)\Big\}\Big\}$$

$$+ h_5 a_5 \left\{\frac{t_1{}^2}{2} + \frac{\theta_5}{6}t_1{}^3\right\} + h_5 a_5 b_5 \left\{\frac{t_1{}^3}{3} + \frac{\theta_5}{8}t_1{}^4\right\}$$

$$+ C_5 \left\{\frac{a_5 \theta_5 t_1{}^2}{2} + \frac{a_5 b_5 \theta_5}{3}t_1{}^3\right\}$$

$$+ S_5 \left\{\frac{a_5}{2}(t_1 - T)^2\right.$$

$$-\frac{a_5 b_5}{2}\left(t_1{}^2 T - \frac{T^3}{3} - \frac{2}{3}t_1{}^3\right)\Big\}\Bigg]$$

$$= 0 \qquad (4.15)$$

Further, for the total cost function $K_{dS}(t_1, T)$ to be convex, the following conditions must be satisfied

$$\frac{\partial^2 K_{dS}(t_1, T)}{\partial t_1{}^2} > 0, \quad \frac{\partial^2 K_{dS}(t_1, T)}{\partial T^2} > 0 \qquad (4.16)$$

And

$$\left(\frac{\partial^2 K_{dS}(t_1, T)}{\partial t_1{}^2}\right)\left(\frac{\partial^2 K_{dS}(t_1, T)}{\partial T^2}\right) - \left(\frac{\partial^2 K_{dS}(t_1, T)}{\partial t_1 \partial T}\right)$$

$$> 0 \qquad (4.17)$$

The second derivatives of the total cost function $K_{dS}(t_1, T)$ are complicated and it is very difficult to prove the convexity mathematically. Thus, the convexity of total cost function has been established graphically. (Figure A)

4.2. Numerical Example

Consider an inventory system with following parametric values. [2]

Crisp Model, A=Rs.200/order, C=Rs.20/unit, h=Rs. 5/unit/year, a=100 units/year, b=0.1units/year, $\theta = 0.01$/year, S=Rs 15 /unit/year.

The solution of crisp model is $K(t_1, T)$ = Rs 404.3429, t_1 =0.7149 year, T = .9639 year.

Fuzzy model,

$$\tilde{a} = (60, 80, 100, 120, 140), \quad \tilde{b} = (0.06, 0.08, 0.10, 0.12, 0.14)$$

$$\tilde{C} = (16, 18, 20, 22, 24), \quad \tilde{S} = (11, 13, 15, 17, 19)$$

$$\tilde{\theta} = (0.006, 0.008, 0.010, 0.012, 0.014), \quad \tilde{h} = (1, 3, 5, 7, 9)$$

The solution of fuzzy model can be determined by following Graded Mean Representation Method.

1. When $\tilde{a}, \tilde{b}, \tilde{C}, \tilde{S}, \tilde{\theta}, \tilde{h}$ all are pentagonal fuzzy numbers.

$$K_{dG}(t_1, T) = Rs.\,414.6096 \quad, t_1 = 0.6908\, year,$$
$$T = 0.9383\, year$$

2. When $\tilde{a}, \tilde{b}, \tilde{C}, \tilde{S}, \tilde{\theta}$ all are pentagonal fuzzy numbers

$$K_{dG}(t_1, T) = Rs.\,406.9852 \quad, t_1 = 0.7135\, year,$$
$$T = 0.9560\, year$$

3. When $\tilde{a}, \tilde{b}, \tilde{C}, \tilde{\theta}$ all are pentagonal fuzzy numbers.

$$K_{dG}(t_1, T) = Rs.\,405.5274 \quad, t_1 = 0.7115\, year,$$
$$T = 0.9596\, year$$

4. When $\tilde{a}, \tilde{b}, \tilde{\theta}$ all are pentagonal fuzzy numbers.

$$K_{dG}(t_1, T) = Rs.\,405.2250 \quad, t_1 = 0.7120\, year,$$
$$T = 0.9603\, year$$

5. When $\tilde{a}\, and\, \tilde{b}$ all are pentagonal fuzzy numbers.

$$K_{dG}(t_1, T) = Rs.\, 404.8978 \quad , t_1 = 0.7131\, year,$$
$$T = 0.9611\, year$$

To show the convexity of cost function $K_{dG}(t_1, T)$, we plot a 3D graph among t_1 and T, where values of both t_1 and T ranging from $t_1 = .65$ to 2 with equal interval,

$T= .84$ to 1 respectively. A three-dimensional graph is shown in the following:

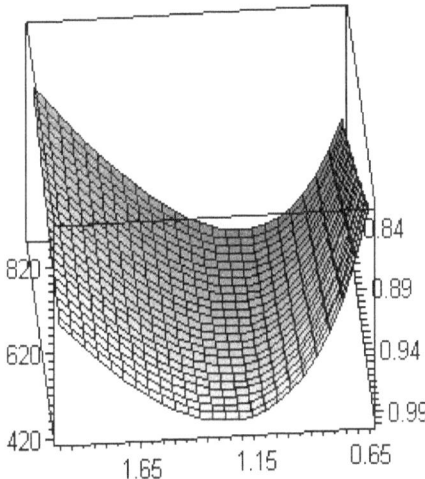

(Figure A) Total fuzzy cost $K_{dG}(t_1, T)$ Vs. t_1 and T.

Consider an inventory system with following parametric values.

Crisp Model, A=Rs.200/order, C=Rs.20/unit, h=Rs. 5/unit/year, a=100 units/year, b=0.1units/year, $\theta = 0.01$/year, S=Rs 15 /unit/year.

The solution of crisp model is $K(t_1, T) = $ Rs 404.3429, t_1 −0.7149 year, $T = .9639$ year.

Fuzzy model,

$\tilde{a} = (60, 80, 100, 120, 140)$, $\tilde{b} = (0.06, 0.08, 0.10, 0.12, 0.14)$

$\tilde{C} = (16, 18, 20, 22, 24)$, $\tilde{S} = (11, 13, 15, 17, 19)$

$\tilde{\theta} = (0.006, 0.008, 0.010, 0.012, 0.014)$, $\tilde{h} = (1, 3, 5, 7, 9)$

The solution of fuzzy model can be determined by following Signed Distance Method.

1. When $\tilde{a}, \tilde{b}, \tilde{C}, \tilde{S}, \tilde{\theta}, \tilde{h}$ all are pentagonal fuzzy numbers.

$$K_{dS}(t_1, T) = Rs.\,419.6059 \quad , t_1 = 0.6797\,year,$$
$$T = 0.9266\,year$$

2. When $\tilde{a}, \tilde{b}, \tilde{C}, \tilde{S}, \tilde{\theta}$ all are pentagonal fuzzy numbers

$$K_{dS}(t_1, T) = Rs.\,408.2810 \quad , t_1 = 0.7135\,year,$$
$$T = 0.9523\,year$$

3. When $\tilde{a}, \tilde{b}, \tilde{C}, \tilde{\theta}$ all are pentagonal fuzzy numbers.

$$K_{dS}(t_1, T) = Rs.\,406.1163 \quad , t_1 = 0.7093\,year,$$
$$T = 0.9576\,year$$

4. When $\tilde{a}, \tilde{b}, \tilde{\theta}$ all are pentagonal fuzzy numbers.

$$K_{dS}(t_1, T) = Rs.\,405.6640 \quad, t_1 = 0.7106\,year,$$
$$T = 0.9587\,year$$

5. When $\tilde{a}$ and $\tilde{b}$ all are pentagonal fuzzy numbers.

$$K_{dS}(t_1, T) = Rs.\,405.1742 \quad, t_1 = 0.7122\,year,$$
$$T = 0.9599\,year$$

To show the convexity of cost function $K_{dS}(t_1, T)$, we plot a 3D graph among t_1 and T, where values of both t_1 and T ranging from t_1 = .65 to 2 with equal interval,

T= .84 to 1 respectively. A three-dimensional graph is shown in the following:

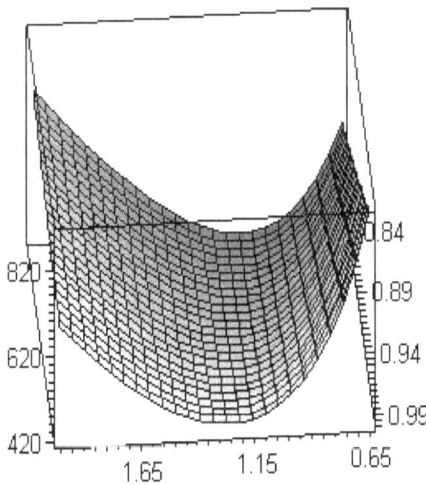

(Figure A) Total fuzzy cost $K_{dS}(t_1, T)$ Vs. t_1 and T.

Chapter-4: CONCLUSIONS

This dissertation presents a fuzzy inventory model for deteriorating items with shortages under fully backlogged condition in which demand is an increasing function of time. Shortages and deterioration are natural in any inventory control system. The proposed model is developed in both the crisp and fuzzy environments. In fuzzy environment, all related inventory parameters are assumed to be pentagonal fuzzy numbers. For defuzzification, graded mean representation and signed distance method is employed to evaluate the optimal time period of positive stock t_1 and total cycle length T which minimizes the total cost. By given numerical example it has been tested that graded mean representation method gives minimum cost.

Future Research Direction

1. The other uncertainties such as supply uncertainty, lead time uncertainty and costs uncertainty can be present in inventory control.
2. Comparing the result of this study with the result of the other studies with different Fuzzy system.
3. Concerning the multi echelons/stages within the supply chain is also interesting for the future research.

REFERENCES

1. "Fuzzy sets as a basis for a theory of possibility," *Fuzzy Sets and Systems* **1**: 3–28

2. C. K. Jaggi, S. Pareek, A. Sharma and Nidhi, "Fuzzy inventory model for deteriorating items with time-varying demand and shortages", *American Journal of Operational Research*, Vol. 2(6), pp.81-92 (2012).

3. D. Dubois and H. Prade (1988) Fuzzy Sets and Systems. Academic Press, New York.

4. D. Dutta and Pavan Kumar, "Fuzzy inventory model without shortage using trapezoidal fuzzy number with sensitivity analysis", *IOSR-Journal of Mathematics*, Vol. 4(3), pp.32- 37. CrossRef, doi: 10.9790/5728-0433237 Nov-Dec (2012).

5. D. Dutta and Pavan Kumar, "Optimal policy for an inventory model without shortages considering fuzziness in demand, holding cost and ordering cost", *International Journal of Advanced Innovation and Research*, Vol. 2(3), pp.320-325 (2013).

6. D. Pertrovic and E. Sweeney, "Fuzzy knowledge-based approach to treating uncertainty in inventory control," *Computer Integrated Manufacturing Systems*, Vol. 7, 1994, pp. 147-152.

7. Fuzzy Logic A Practical Approach by F. Martin McNeill and Ellen Thro.

8. H. C. Chang, "An application of fuzzy sets theory to the EOQ model with imperfect quality items," *Computers and Operations Research*, Vol. 31, 2004, pp. 2079-2092.

9. H. C. Chang, J. S. Yao, and L. Y. Ouyang, "Fuzzy mixture inventory model involving fuzzy random variable lead-time

demand and fuzzy total demand," *European Journal of Operational Research*, Vol. 169, 2006, pp. 65-80.

10. Hass. Michael., (2009), *Applied Fuzzy Arithmetic*, Springer International Edition, ISBN 978-81-8489-300.

11. http://en.wikipedia.org/wiki/Fuzzy%20concept?oldid=638161281.

12. J. S. Yao, S. C. Chang, and J. S. Su, "Fuzzy inventory without backorder for fuzzy quantity and fuzzy total demand quantity," *Computers and Operations Research*, Vol. A FUZZY INVENTORY SYSTEM UNDER SUPPLIER CREDITS 251 27, 2000, pp. 935-962.

13. jershan chiang, jing-shing yao, and huey-ming lee," **Fuzzy Inventory with Backorder Defuzzification by Signed Distance Method"** JOURNAL OF INFORMATION SCIENCE AND ENGINEERING 21, 673-694 (2005).

14. Klaua, D. (1965) Über einen Ansatz zur mehrwertigen Mengenlehre. Monatsb. Deutsch. Akad. Wiss. Berlin 7, 859–876. A recent in-depth analysis of this paper has been provided by Gottwald, S. (2010). "An early approach toward graded identity and graded membership in set theory". *Fuzzy Sets and Systems* **161** (18): 2369–2379. doi:10.1016/j.fss.2009.12.005.

15. L. A. Zadeh (1965) "Fuzzy sets". *Information and Control* 8 (3) 338–353.

16. Lily R. Liang, Shiyong Lu, Xuena Wang, Yi Lu, Vinay Mandal, Dorrelyn Patacsil, and Deepak Kumar, "FMtest: A Fuzzy-Set-Theory-Based Approach to Differential Gene Expression Data Analysis", BMC Bioinformatics, 7 (Suppl 4): S7. 2006.

17. M. J. Yao, P. T. Chang, and S. F. Huang, "On the economic lot scheduling problem with fuzzy demands," *International Journal of Operations Research*, Vol. 2, 2005, pp. 58-71.

18. M. K. Maiti, "Fuzzy inventory model with two warehouses under possibility measure on fuzzy goal," *European Journal of Operational Research*, Vol. 188, 2008, pp.746-774.

19. P M Pu and Y M Liu, "Fuzzy Topology1, neighborhood structure of a fuzzy point and Moore- Smith Convergen ce", *Journal of Mathematical Analysis and Application*, Vol. 76, pp. 571-599, (1980).

20. Richard Dietz & Sebastiano Moruzzi (eds.), Cuts and clouds. Vagueness, Its Nature, and Its Logic. Oxford University Press, 2009

21. Ruxian Li1, Hongjie Lan1*, John R. Mawhinney" **A Review on Deteriorating Inventory Study"** *J. Service Science & Management*, **2010, 3: 117-129** doi:10.4236/jssm.2010.31015 Published Online March 2010 (http://www.SciRP.org/journal/jssm).

22. S. H. Chen, S. T. Wang, and S. M. Chang, "Optimization of fuzzy production inventory model with repairable defective products under crisp or fuzzy production quantity," *International Journal of Operations Research*, Vol. 2, 2005, pp. 31-37.

23. Sandeep Mehan & Vandana Sharma, "Development of traffic light control system based on fuzzy logic". ACAI '11 Proceedings of the International Conference on Advances in Computing and Artificial Intelligence 2011, pp. 162-165.

24. Susan Haack notes that Stanisław Jaśkowski provided axiomatizations of many-valued logics in: Jaśkowski, "On the rules of supposition in formal logic. *Studia Logica* No. 1, 1934. See Susan Haack, *Philosophy of Logics*. Cambridge University Press, 1978, p. 205.

25. Susan Haack, Deviant logic, fuzzy logic: beyond the formalism. Chicago: University of Chicago Press, 1996.

26. Susan Haack, *Philosophy of Logics*. Cambridge University Press, 1978, p. 165.

27. Tanthatemee T., Phruksaphanrat B., *Member, IAENG,"* Fuzzy Inventory Control System for Uncertain Demand and Supply",proceeding of the International multiconference of engineering and computer scientists 2012Vol II,IMECS2012,march 14-16,2012,hongkong.

28. www.mathsisfun.com/definitions/Pentagonal-number.html.

---------------------------------------xxxxx---

www.ingramcontent.com/pod-product-compliance
Lightning Source LLC
Chambersburg PA
CBHW062123220526
45471CB00010B/3858